Life RAKED IN

Penned in a Wild Blueberry Field in Maine

Written and Illustrated by
Gail J. VanWart

Life Raked In

Written, Illustrated and Designed by
Gail J. VanWart

First Edition
© Copyright Gail J. VanWart 2011

Second Edition
© Copyright Gail J. VanWart 2017
All Rights Reserved

Published by
Out of the Blue, LLC
PO Box 102, Holden, ME 04429
outofthebluellc.com
at
Peaked Mountain Farm
and Native Pollinator Sanctuary
Dedham, Maine
peakedmountainfarm.biz

LCCN 2011919401
ISBN 978-0-9848206-5-8

*A harvest of poetry, recipes and thoughts
raked in from the Wild Blueberry fields of life.*

Thanks to my family and Nature, herself,
for providing me with the thoughts I have
transformed into these words—even if they
never realized they were doing so.

Life RAKED IN

Questions from an Unknown Poet

Do I, the miniature
among giants,
dare to soil these pages
with inklings
of my thoughts,
and pawn them off
as poetry?

How can I compare
the whispers of my soul
to great collections
of words
already immortalized
with fame?

Yet, who am I
to deny
this world,
which has given me
everything I own,
even one syllable
of what any person
cares to say?

Questions from an Unknown Poet
was originally published in the1988 Edition of
Poetic Voices of America
Sparrowgrass Poetry Forum Inc.

Introduction

I reside with my husband, Daniel, at the foot of Peaked Mountain in Dedham, Maine, overlooking the Wild Blueberry fields that have been a part of my life since birth. As a fourth generation steward of our farm, I credit it for teaching me some of life's most important lessons and inspiring creativity in everything I attempt. After raising three awesome sons, I presently enjoy a much less demanding form of parenting as a grandmother. It's my feeling, now is the time to publish some of the piles of clutter I've accumulated over the course my lifetime so that, when it is my turn, I can die in peace. Hopfully, you'll enjoy this second edition of my first official collection of poetry and a few of my favorite Wild Blueberry recipes that were, until 2011, just piles of my clutter.

– Gail J. VanWart

Photo © Daniel VanWart 2009 ~ Gail with her faithful dog, Blae.

New Bloom

Every cycle of life starts in its own spring with the birth of life's new blooms. Springtime in Maine is filled with a buzzing of love songs the bees are singing across the state as they pollinate the Wild Blueberry fields, which have been a part of Maine's natural environment for over 10,000 years. It reminds us of the very nature of things and how the past is woven so intricately into the fabric of our future.

Simply Great

A little boy
and his dog
are playing
in the sun.
The little boy
throws a stick
to make his puppy run.
As I watch them
from afar,
I feel so very glad
for the simple things
in life
is a great gift
they have.

Simply Great
was originally published in the
1988 Edition of
American Poetry Annual
The Amherst Society

1

I Want to Paint You Rainbows

Are pictures truly worth
a thousand words?
If they are,
I'd like to paint for you,
the many things
left hidden in my heart,
unexpressed,
by saying "I love you".

Words can't full explain
sunshine in the rain —
phrases can't reveal
all the things I feel.

If I could find a way
to paint you rainbows
in the sky,
maybe then I could describe
my love for you
words hide.

I want to paint you rainbows,
I want to draw you sunshine,
I want to color out the rain,
I want to portray
my love for you.

Peaked Mountain Farm's
Wild Blueberry Conserve

Makes 4 Half Pints

4 Cups Sugar
2 Cups Water
1/2 Lemon *(thinly sliced)*
1/2 Orange *(thinly sliced)*
1 Apple *(cored, peeled and chopped)*
1 Quart Wild Blueberries

Mix sugar and water in saucepan and bring to a boil. Add lemon, orange and raisins. Simmer 5 minutes. Stir in blueberries and cook mixture rapidly. Stir frequently as mixture thickens. *(Add favorite spices to make it your own!)*

When almost to gelling point, ladle hot conserve mixture into hot sterilized jars, leaving 1/4 inch headspace. Place sterilized caps on jars and process by immersing them into a canning pot of boiling water for 15 minutes. Remove jars from boiling water bath and let sit on a towel on a counter until cool. Check covers to see if the caps are vacuum sealed. If a jar lid does not seal properly, store jar in refrigerator and use it right away.

The Treasure
in the Attic

The bandanna
was tied
securely in place,
as my colorful
headdress of war.
I then proceeded,
armed heavily
with broom,
to tackle
the deadly chore.

Though spectral images
resented my raid,
I marched bravely
into the attic,
where a decade
had built
its strong barricade
of historic piles
of havoc.

In that dingy
catch-all vault
the dust shrouded past
declared
it would attack
my memory
and drive me out of there.

Even webs of neglect
stretched ear to ear
like an evil grin
on a pitiful,
hairless doll.

Surely,
just a tactful ploy
to ultimately
make me stall.

Against all doom
I fought the gloom!
Determined
to trash
the junk
I'd stashed.

In my handy garbage bag
what was a dress
became a rag.
Into that bag
I swiftly threw
a cracked croquet ball
and worn out shoe.

I'll not falter,
I'll not sway,
till all this trash
is thrown away.
This time
I'll clean it all,
damn attic will sparkle,
wall to wall!

I thought...

Then I found it,
in a dusty heap
I was vengefully
about to sweep.

4

Suddenly,
the patter
of little feet
romped playfully
in my mind,
so sweet,
they
tickled
memories
of that spring day
you came into
my heart to stay.

I dusted
its tiny buckle
and gazed
at its smallness
in my hand.
How its significance
could be
discarded,
I couldn't
understand.

Downstairs,
you faithfully awaited
the moment of my return.
Patiently enduring
a lack of
company you yearned.
Ten years
have changed your image,
from flippant puppy
to loyal friend.
How could I fail
to remember
the treasure
you have been?

I hung the puppy collar
on a
newly dusted hook.
I gathered up
my half filled bag,
and left
my slightly tidy nook.

Her tail
swished about me.
Joyfully,
when I reached
the hall.
Today wasn't meant
for cleaning,
it was designed
for playing ball.

I sensed my attic
chuckle,
as we frolicked
in the yard.
It thinks I lost
the battle,
but I gave it
no regard.

Time is eternal
for an attic,
not
for a little dog.

Dedicated tothe memory of
Preshus, the Border Collie
who loved Wild Blueberries

Pitiful Worm

You wiggle and squirm,
you pitiful worm.
Loved by so few
you hide from our view,
though your value is known
to the Robin,
fishermen,
and curious boys.

You're a feast,
a tool,
a toy.
But you've never caused
tremendous joy,
nor pain.

You are so ugly to behold,
so repulsive to touch;
poor little worm,
I pity you so much.
In your hole
down in the ground,
how dreadful it must sound
to have the world
clomping over you.

Of Trickery

Sunlight trickling through the pane
flickers shadows on the floor,
of trees which gaily dance inside
though firmly rooted out of doors.

These silhouettes cease to live
as the world moves silently on,
letting sly reflections give
impressions of a traveling sun.

Rarely things are as appears:
reality apes deception.
Sadness can't be judged by tears,
nor knowledge by perception.

Be wary of how you hear,
skeptical of what you see,
until you feel the earth in motion
and as steadfast as a tree.

Of Trickery
was originally published in
the 1988 Edition of
American Poetry Annual
The Amherst Society

A Natural Friend

A tree bends lovingly
with the wind,
offers branches
to nest within,
opens hollows
in which to hide.
affords shade
to a passerby.
A tree bends lovingly
as it extends
open arms
to many friends.

The Nature of Things

The wind cries hysterically
as sun rays seer the earth,
and tides thunder in appall
as it dumps
remnants of better times
on earth's decaying shores.

Funny how these elements
remain so strong,
in spite of our best efforts
to destroy our natural habitat.

Stars

Stars are places
in the sky
we can barely see.
Why I wish upon one,
is really beyond me.

Yet, I can't help but
wonder if
someone,
out there
in space,
is wishing back at me.

WILD & FROSTY

Makes four 10 oz. servings.

1 1/3 Cups Wild Blueberries
(fresh or fresh frozen)
1 Can Frozen Lemonade Concentrate *(6 oz.)*
1 Pint Vanilla Ice Cream
1 Bottle of Ginger Ale *(20 oz.)*

Spoon 1/4 of the lemonade concentrate into
four 10 oz. tall glasses. Add 1/3 cup blueberries
(slightly crushed) and a large scoop of vanilla ice cream
to each glass. Fill with ginger ale and mix. Serve with
straws and long handled spoons.

Why Do I Feel?

Why do I feel
the woes of the world?
I curse wrongs
I cannot right,
that wake me in the
dead of night—
and I shutter
to think
I'm only me.

I flinch with pains
of the starving,
the sick,
the dying.

I vomit
from thoughts
of it being
somebody's child,
somebody's parent,
somebody's loved one—
friend.
Praising God
when it's not "mine"
doesn't ease my pain
for someone else,
somewhere.
But, I'm only
me.

I've died
a thousand times
on planes,
on trains,
in automobiles.

I've drowned
in lakes,
rivers,
and oceans.
I've screamed
as I've fallen
from tall buildings
and leaped
off bridges
or woke up in the
recovery room
with but half my limbs.

I've choked
my last breath
when the fingers of
the strangler's hands
tightened.
I've hidden in shame
with the sperm of the rapist
running down my legs.

I've known death's bullet,
a knife piercing my heart,
and fire devouring my flesh.

I've frozen to death,
and I've cried —
sobbing for the woes of
the world
that invade my soul,
my heart,
my mind —
weeping because
I'm not a God
and will never be
more than
me.

I shout in despair,
"The woes of the world
are too much to bear
or *even* care about."

Then I feel
the flowers blooming,
fresh drops of
a rain shower,
sun rays,
rainbows forming,
sunrises dawning,
sunsets.

People smiling,
dancing,
skating,
singing,

Bees humming,
birds mating,
cotton candy clouds.

Friends laughing,
lovers embracing,
babies nursing,
puppies playing,
Pine trees swaying,
grandparents with photos
to proudly display.

I feel helping hands
extended,
broken bodies mended,
and hope.

I feel others
caring,
sharing each burden
I'm bearing,
wiping away my fear,
absorbing
my tears of woe.

And I weep with joy
realizing
I'm only me,
but feeling
I'm not alone.

As Wild Blueberry blossoms transform into their fruit, the struggle of growth begins. Challenges are ever present in the Wild Blueberry fields, as they are in our daily journey from infancy to maturity. Growth is a time of coping, adapting, and learning how to overcome whatever tries to hinder it.

A Long Way to Grow

Children grow up
each time they fall down.
Each sun passing over them
is another day under them.
Strange, that I've found,
as I stumble to the ground,
I am but a child
at the age of sixty-two.

The Wind and I

Soft is the wind
that caresses the grass
with gentle kisses
as it breezes past.

Strong is its will
to rule land and sea,
with conquering gales
it sweeps
eternity.

So quick is its spirit
you notice not its form,
you only see its pathway
of peaceful calm or storm.

Wherever it is,
wherever it goes,
brushing the sands
or gliding the snows,
the wind and I are friends,
passing time of day,
never knowing where we're going
yet, forever on our way.

The Wind and I
was originally published in
the1988 Edition of
A Time to be Free, Volume II
Quill Books

To a Mother, From a Mother

I'm ironing in the morning,
washing clothes at noon,
scrubbing dirty dishes
and a million forks and spoons.
They say woman's work is never done,
but mine is always just begun!
When at night I fall into bed
even busy dreams fill my head.

Children scatter themselves
here and there,
with kittens, cats, and teddy bears.
The yard is filled with bikes and dogs,
and dirty pockets
hold slow witted frogs.
What on earth am I to do
to control this homemade zoo?
Guess I'll do what I can today
with chores undone from yesterday.

There is a time, now and then,
I think my life a bore,
and no greater will it be
than a daily household chore.
There's not much time
for a mother's dreams
of seeking notice in this world,
or doing all the marvelous things
she'd dreamed of as a girl.

Yes, I'm *only* a mother
and I'll hardly be thanked
for the smart little ego
I just had to spank.
Not destined to be famous,
I alone shall pride my name.
My talents may seem aimless
and my poetry all in vain.

To all other mothers
understanding
what I've said;
don't let it get you down,
stop and think ahead...
to when your children
may have children
they'll be glad to tuck in bed!
Just when you've begun to frown,
you just might smile instead.

So, if nobody wants
to hear your poems
or look upon your art,
the fact that you're a mother
sets your talents far apart.
Cause, if you're ever able
to find time to let it out,
you know
you'll be dressed in sable
and *your* name
this world will shout.

My Nobody, Somebody, Everybody

My husband says he's "nobody"
when nobody takes out the trash,
only to be "somebody" when
somebody speaks too rash,
and multiplies to "everybody" if
everybody does things wrong...
he feels like he's an entire crowd
where only one of him belongs.

Perhaps he doesn't realize,
although just one of him exists,
he can be "anybody"
my little heart insists;
for "nobody" makes me happier
than the "somebody" who cares for me,
he's "everybody" I need in life...
he's all the world to me.

My Nobody, Somebody, Everybody
is dedicated to my husband, Dan, and
was originally published in the1990 Edition of
Great Poems of the Western World - Vol. II
World of Poetry Press

Dan's Wild Blueberry Crunch

Combine Base:
1 Pint Wild Blueberries *(fresh or fresh frozen)*
2 Tablespoons Flour
1/4 Teaspoon Salt
1/2 Cup Sugar
2 Tablespoons Lemon Juice

Spread into a well greased nine-inch pie pan.

Top With:
1 Cup Flour
1 Cup Quick-Cooking Oatmeal
1/2 Cup Brown Sugar (firmly packed)
1/2 Teaspoon Salt
1/2 Teaspoon Vanilla Extract
1/2 Cup Butter

Cut the 1/2 cup butter into combined topping ingredients until it resembles coarse meal. Sprinkle on top of blueberry mixture.

Bake at 350° for 30-40 minutes.

Safe Containment

I was alone in the kitchen
when it challenged me.
The struggle which resulted
was hideous to see.

My left hand clenched it firmly,
while the other gripped its neck.
Though I tried to twist it sternly,
its strength would not relent.

It never budged a fraction,
and I could clearly see
a need for stronger action,
before it got the best of me.

So, I slapped it on the bottom,
I banged it with my fist,
then grabbed it by the neck again,
and twisted firmly with my wrist.

I grunted,
I groaned,
I cursed it,
and I moaned —
it mattered not what I wailed,
my dominance did not prevail.
With shattered nerves
now on the brink,
my patience was out of order,
when I shoved the culprit
into the sink
and scalded it
with hot water!

The mindless thing
paid this no heed,
vacant of all care.
This fueled the sparks
of rage in me,
into a frightful flare.

I threw a towel around it,
and strangled with all my might,
yet, the gall-dang thing
still ignored
the violence of my fight.

So, I grabbed it with less mercy,
for resisting each encounter.
I picked it up
then firmly
bashed it on the counter.

When that assault did no good,
I broke all golden rules.
I reached into the whatnot drawer
and confronted it with tools.

Admittedly, it wasn't nice
to attack it with a vice,
yet, even that weapon made me sigh,
try, after try, after try.

My brow was sweated
as my arms grew weak,
but the rage inside me
was at its peak.

The damn thing coolly gave me
such a carefree, glassy glare,
it further engaged my tongue
into language that is rare.

"I'll fix you,"
I wryly hissed.
"You'll be darn sorry
you made me pissed!"

I turned once more
to the whatnot drawer,
and swiftly drew out a hammer.
With one powerful,
deadly blow
I ended all further clamor.

Spaghetti sauce spewed
down the counter
to the floor.
It sprayed the walls,
and the oven door.
Glass bits scattered,
near and far.
By God, I'd opened
that cursed jar!

Not that it really mattered now —
and I solemnly wondered how
mankind could *ever* enjoy
an easy meal,
since it's invented
the damn safety seal.

Mother's Thoughts

When the distant patter of little feet
run through her memory,
of distant times in days gone by
which only memories can see,
she pauses from her daily chores
for a little while...
and she alone appreciates
the reasons for her smile.

Since seasons only come to go,
there's nothing else she can do...
when there is no more Goldie Locks,
no more Little Boy Blue...
but to reminisce from time to time
when she thinks of you.
Wouldn't it be nice
if she was sure
you thought about her, too?

Mother's Thoughts
was originally published in the
1991 Edition of
World of Poetry Anthology
World of Poetry Press

I Can Only Hope

Yes, yes...I know.
You're *not* my baby anymore.
In fact,
I know you're
almost a man,
who'll soon depart my door.
So, you ask me why
I even bother to try
to bring you up anymore.
You feel you're treated
like a helpless pup...
and wish that I'd
just give up.

Yes, I know you think
I'm your greatest foe on earth,
and that everything I say to you
is nagging without worth...
even though I've never taught you
wrongs,
or scolded you for doing right.
I've never offered you harmful
drugs
or challenged you to a fight.

No, you won't be a child
much longer,
and when your mind
is stronger
you will understand,
why I don't say to hell with it
and wash you off my hands.

You see,
even if I can not make you
practice what I preach,
or even listen to the lessons
I constantly try to teach...
I can still hope
that you will learn enough.

I can hope that you will be
just bright enough
to win a battle,
without your fist,
and secure enough
to laugh at jest.
For the day that I can see
you are just good enough
to breed a little integrity
into the human race,
and are capable
of forgiving grace,
I'll unleash you.

I won't expect you
to be perfect
when you depart...
I can only hope you'll be
just perfect enough
not to break your own heart
when you find yourself
still a puppy
in a dog eat dog
world.

Gail's Wild Blueberry Gingerbread

This dessert is always a hit!

2 Cups Flour
1/2 Cup Shortening
1/2 Teaspoon Salt
1 Cup Sugar
1 Egg
1/2 Teaspoon Ginger
1 Teaspoon Cinnamon
1 teaspoon Baking Soda
3 Tablespoons Molasses
1 Cup Sour Milk*
1 Cup Wild Blueberries *(Fresh or Frozen)*
3 Tablespoons Sugar

3 Tablespoons Sugar
to sprinkle on top before baking

Confectioner's Sugar
to sprinkle on top after baking

1/2 Cup Fresh Wild Blueberries
for garnish (if available)

Cream shortening and add salt and sugar gradually. Add unbeaten egg and beat until light and fluffy. In separate bowl, mix flour, ginger, and cinnamon. In another bowl mix baking soda with sour milk, stir until soda is dissolved. Then add dry ingredients and sour milk mixture alternately to the creamed mixture. Add the molasses and fold in the cup of wild blueberries. Turn into a greased and floured pan (either a 9" x 9" square or 10" round pan works well). Sprinkle top with 3 tablespoons sugar and bake at 350° for 45-60 minutes. When done, turn onto rack to cool. When cool, place on a plate and sprinkle with confectioner's sugar and garnish with fresh blueberries.

* If you don't have sour milk on hand, add 2 tablespoons vinegar to one cup of sweet milk.

Peachy Clean

My washer plainly tells me
which buttons I should push
for colors, whites,
or delicate fabrics
I wear around my tush.
In this day of automation
and special fabric
bleach,
I can't figure
for the life of me
what I do
that turns
the white clothes
peach.
Being of female gender,
this personally
I can bear.
What kills me
are the daggers
my three sons
do glare,
as they reluctantly
dress their bodies
in peachy clean
underwear.

Lily Pad World

Sometimes,
the world seems like a Lily Pad
floating in space
with an invisible umbilical cord
keeping us in our place.

Though we toss
and turn
through currents of time,
we stay here
attached to the unknown.

Sometimes,
we wish we were frogs
able to jump off
to sit on
another's
Lily Pad world.

Change

Where is my poetry?
Where has it gone?
Where are those midnights
and early, early dawns?
Have I forgotten
the words to my song;
slipped on shoes
to alienate the grass
and shed my jeans
for designer pants?

Peaked Mountain Farm's Chilled Wild Blueberry Soup

Soup:

    3 Cups Wild Blueberries
      (Fresh or Frozen)
    2 Cups Apples *(Peeled and Cored)*
    1/3 Cup Lemon Juice
    3/4 Cup Apple Juice
    1/4 Cup Honey
    2 Cups Plain Low-fat Yogurt
    3/4 Cup Light Sour Cream

Puree wild blueberries, apples, apple juice, and honey in a blender. Add yogurt and sour cream. Blend for a few seconds and then chill. Serve with a dollop of sour cream topping (below).

Topping:

    3/4 Cup Light Sour Cream
    3 Tablespoons Horseradish njjjk
    1 1/2 Teaspoons Lemon Juice

Mix sour cream, horseradish, and lemon juice together.

Planting Seed

I planted
a seed today.
Maybe it will grow.
I'll have to wait to see,
that's the only way
I'll know.

Maine's Fabulous
Wild Blueberry Cake

Serves eight.

2 Eggs
1 Cup Sugar + 3 Tablespoons to sprinkle on top
1/2 Teaspoon Salt
1/2 Cup Shortening
1 Teaspoon Vanilla
1 1/2 Cups Flour
1 Teaspoon Baking Powder
1/2 Cup Milk
1 1/2 Cups Fresh Wild Blueberries *(coat blueberries lightly with small amount of the flour)*

Separate eggs carefully, beat egg whites until stiff and blend in 1/4 cup of the sugar to keep them firm. In another bowl, cream shortening with salt, vanilla, and remaining 3/4 cup of sugar. Add unbeaten egg yokes and beat mixture until light and creamy. Mix dry ingredients together and add alternately with milk to the creamed mixture. Then fold in the egg whites and the flour coated blueberries. Turn into an 9" x 9" pan that has been coated with shortening and lightly dusted with flour, then sprinkle top lightly with sugar and bake at 350° for 45-60 minutes. Serve either warm or cool.

Top with whipped cream if you so desire.

Maine's Day

Faithfully toiling
from the crack of spring's dawn,
bestowing new life
all through your morn.
Flowering the meadows
all summer's noon,
and rewarding us with harvest
in your late afternoon.

Only after painting every leaf
for the spectacular sunset
of autumns eve,
will you rest
under a white blanket
of winter's snow.

Maine,
your seasons four
outside my door
are ever changing
in a way
as if they're
but one day
in your own
dimension
of time.

Fast Act
Wild Blueberry Sorbet

Easy to make behind the scenes
ahead of time. Serves six.

4 Cups Wild Blueberries (fresh or frozen)
1 Can Apple Juice Concentrate (6 oz.)

Combine Wild Blueberries and Apple concentrate in
a blender to liquefy. Pour into an 11" x 7" baking pan
and cover. Freeze for about 2 hours until firm around
the edges. Chop up the semi-frozen mixture and return
to blender and blend until smooth and completely
melted. Then spoon into a 9" x 5" loaf pan and freeze
again until firm. Best if served within a couple of days.

What is Who?

What is what and who is who?
I'm not sure I know.
Do you?
One can be one
or actually mean two,
I can be "me"
and I can be "you"...
it all depends
on the point of view.

Real Cool and Easy
Wild Blueberry Pie

1 Graham Cracker Pie Crust
  (Buy pre-made or make your own)
1 Cup Sweetened Condensed Milk
1/2 Cup Lemon Juice
1 Pint Fresh Wild Blueberries
8 Oz. Tub of Whipped Topping
1/3 Cup Wild Blueberries for Garnish

1. Stir milk and lemon juice together until well mixed.

2. Mix in the pint of clean fresh blueberries. Fold in the whipped topping. Spoon into pie crust.

3. Freeze for 5 hours or until firm.

4. Let stand at room temperature 30 minutes before serving. Garnish with 1/3 cup blueberries.

Graham Cracker Nine-Inch
Pie shell

1 1/4 Cups Graham Cracker Crumbs
Roll about 16 graham crackers with rolling pin.

2 Tablespoons Sugar

1/3 Cup Melted Butter *(Add more if needed)*

Mix ingredients together and press firmly into a pie plate.

NOTE: You can also make this recipe without the sugar.

Harvest

We never know at the beginning of a crop cycle what the harvest will yield. The final tally can only come after all the Wild Blueberries have been raked in. Sometimes we're happy with the numbers, other years we aren't. But, as one harvest ends, we file its data away to refer to as we look beyond it and start to prepare for the next cycle, applying lessons learned from the past to the prospects of a future crop.

My Memories

Rumblings fill my head,
as I lay tossing on the bed,
night has smothered out the day,
but, its memories long to stay.
I file away each event,
carefully, least I forget...
the smiling of a small boy's face,
the hallowed glow of love's embrace.
Each new memory unlocks a door
to infinite treasures of times before.
Though the days may choose to part,
I cherish their essence within my heart.
Softly now, I drift into sleep,
sated with the mementoes
I'll always keep.

My Memories
was originally published in
the 1988 Edition of
On the Threshold of a Dream
The National Library of Poetry

Grandma's Rocker

Squeak, squawk,
the rocker rocks,
in time to
click-clicking,
of needles knitting,
and causes
its shadow to dance
to and fro.
First it shrinks,
then it grows,
thin and tall,
fat and low,
to and fro,
to and fro...
nobody notices
for nobody's there,
while she sits and knits
in the old rocking chair.
Nobody sees the shadow
dance,
or hears the melody
of an entire lifetime
being knit
from her memory.
Her fingers
know the stitches,
all too very well,
which her eyes
now struggle to see.
Knit two, pearl one,
bind off three...
the pattern flows
unconsciously.

She follows
its monotony,
rocking
to and fro,
clicking
through another row.
For time
after time,
day
after day,
she'd amused
herself this way...
at least
she'd spent
the time.
Sunlight floods
her window
and gently drenches
her aged hands.
It splashes
little, speckled spots,
from her golden
wedding band.
They pounce about
the flowers,
on the faded
papered wall,
but she continues
knitting,
never heeding
them at all...
she's busy
entertaining

company,
who's paid
her mind
a call.
He embraces her warmly
in long ago,
as she rhythmically rocks,
to and fro...
knitting memories
of yesterday,
into the booties
of youths' today.
Her wrinkled hands
are young again,
her womanhood
is new,
as he gently
holds her hand,
and softly says,
"I do."

Sixty-seven years
magically
disappears
into hills and valleys
far away.
Crossing bridges
of sunny days,
which span the rivers
of stormy weather,
and enters
their life's time together...
where eleven tiny babes
were born,

three of which
they'd always mourn.
Yet together
they go on
and on...
planting their fields
with dreams,
each spring,
then settling for
what the harvest brings.
Again
and again...
until
she silently walks on
down life's road,
while he lingers forever
at its bend.

Rocking slowly,
to and fro...
she finishes
the final row,
and quietly dries
two lonely tears,
she'd controlled
for forty years.

Today's booties
slip
to the floor,
and Grandma's rocker
rocks
no more.

Wall Street Blues

They gather
like clusters
of blueberries
on a stem,
bunching together
to form
a healthy
array
of diverse
varieties
in the field.

As they grow,
their rhizomes
spread
across
the land,
reaching
for nourishment.
to feed
their existence.

Another crowd,
resembling
foxes
watching
a hen house,
feasts on
harvests
stolen
from
the
fields,

thinking
they control
this nation's crops
and yields.

Yet, the crop,
newly planted,
but
firmly rooted,
on Wall Street,
feeds upon
a far more sustainable
and organic
fertilizer
in which to
grow
a nation
of equality.

Whether
you are left,
right,
independent,
or wrong,
you
can not
deny
our nation
is singing
for a
just
harvest.

The Common Death

She was born
from dark beginnings,
innocent and unaware
of the price
her life would bear.
Without choice
or question
she entered a place
where reason has
no rhyme,
in this devious,
uncontrollable
entity,
called time…
And now she's gone.

Lacking
egotistical volume,
she modestly
lies inside her grave,
without
worldly recognition
or valor
among the brave.
No Purple Heart
hails
her battles,
in her wars of life.
No trophies
for acomplishments,
or honor
for her strife.
She simply marched
dutifully
into the hereafter,
barren of all
mention
of her tears
and laughter.

Forgotten are the
wondrous deeds
meticulously
chiseled into leaves
on her individual
tree of life.
She is now
just a remnant of
a mother,
and a wife.

No more.
She's just plain
no more.
No more than an eroding
marble headstone
to mark her spot
of decaying bone.
No more than
an inscription
that tells her name,
but nothing of her
lack of fame.
No more than the
puny potted Geranium
left on a Memorial Day,
which also passes away,
in time.

Or worse,
no more than that wreath
of phony flowers,
so weather faded,
it degrades
the only
lingering
marble mockery
of her life
remembered.

Quick and Nutritious
Wild Blueberry
and Honey Smoothie

Makes enough for two.

1 Cup Wild Blueberries *(fresh or frozen)*
1 Vanilla Yogurt *(6 oz,)*
1 Tablespoon Honey
1/2 Cup Ice *(about 3 Cubes)*

Pour ingredients into blender and mix
at high speed until smooth.

Serve immediately.

American Pride?

After passing the poison,
take the pill,
build mighty necessities
to help you kill,
drench the earth in acid rain,
feed our children GMO grain,
sooth their ulcers with plastic paste,
bury their bones in nuclear waste,
swell in pride,
but admit at last,
the human era
is in grave danger
of being
a fad
of the past.

Crossroads

Having seem life from all directions,
I now know
which road to take at the crossing.
The road to happiness
is winding and
traveled less.
The road to sorrow
is straight and wide
to accommodate all its travelers.

Balance

The scales are tilted;
life is not fair.
There's too much heartache
and not enough care.
I try to measure
my life's worth,
but can only weigh
my dreams.

Repeating Chain

It rains in the meadow,
it rains on the trees,
it rains on the mountains,
it rains on me.

It shines on the meadow,
it shines on the trees,
it shines on the mountains,
it even shines on me.

Rain is the drink
which quenches all that thirst.
The sun dries things up again
as it warms the earth.

As it shines on everything,
you will again long for the rain.
That's when you understand
life is a repeating chain.

End of the Cycle

All cycles must complete one full circle of life before another cycle can begin. In the Wild Blueberry fields, we complete it with the pruning of the plants that were harvested so they can again bloom, grow fruit, and provide a healthy harvest. And, so it is with everything.

My Epitaph

Look deep
within the mirrors of your mind;
that's where I will be,
among the soft reflections
of the days you spent with me,
for when you remember
that I was,
I will always be —
deep within the mirrors
of your mind.

Sweet Immortal Poem

Song of the soul,
treasure of the heart,
flow gracefully from the mind
in colorful imitation
of rainbows,
drenched in the sun and rain
of life.

Lift up
on gentle rhythms of rhyme,
arching
your pages of thought
to reflect
your inclinations
into prisms
of awareness
to be reflected,
again and again,
throughout the sweet immortal
chapters of time.

Sweet Immortal Poem
was originally published in
the 1988 Edition of
**The National Poetry
Anthology** by
New York Poetry Anthology, Inc.

Best Ever You Network

Blue Ribbon
Award Winner

*"Gail VanWart is Maine's Best
Blueberry Poet. A delightful and
heartfelt collection of poems and
recipes that is sure to be a
family favorite."*

~

Elizabeth Hamilton-Guarino

**Founder/CEO/LifeCoach/Author
The Best Ever You Network
Magazine.Radio.Web.Community**

besteveryou.com
beynetwork.com